Game Theory:

A Simple Introduction

Also by K.H. Erickson

Simple Introductions

Choice Theory
Financial Economics
Game Theory
Game Theory for Business
Investment Appraisal
Microeconomics

Game Theory:

A Simple Introduction

K.H. Erickson

© 2013 K.H. Erickson

All rights reserved.

No part of this publication may be reproduced, stored in or introduced into a retrieval system, or transmitted in any form or by any means, including electronic, mechanical, photocopying, recording or otherwise, without the prior permission of the author.

Contents

Introduction	6
A Game Matrix	9
The Prisoners' Dilemma	12
The Free Rider Problem	21
Strategic Codependence	29
Repeated Games and Tit for Tat	35
Backward Induction	41
Cartels	49
Asymmetric Information	56
Marketing	66
Fight or Flight	74
Zero-Sum Games	84
When Game Theory Fails	90

Introduction

What exactly is game theory and why is it important? It is a tool used to model interactive decision making, and it shows what's going on in people's minds as they interact. It's used in Mathematics, Economics, Business, Politics, Biology and other disciplines to attempt to make sense of what may at first appear to be irrational behaviour or unexpected outcomes. Game theory suggests that people can always be relied upon to follow their individual incentives and will expect others to do the same, although these incentives will not always lead to the best results. It can offer an explanation as to why people may engage in self-destructive behaviour, consistently achieve worse outcomes than expected, or achieve great results with very little effort.

In simple terms the idea is that life itself is a game and that you and everyone else are playing one hundred percent of the time, whether you want to be or not. It argues that every interaction between two living things sees each make a choice, and that the decision they make depends entirely upon the payoffs or level of gains on offer. Everyone is a player and they're always trying to stay one step ahead of the competition, using mind games if required. Whether it's an argument between two lovers, the boss of a company and the job candidate he interviews, a worker deciding how much effort to exert, or a buyer who bids in the last seconds of an auction, they're all playing the game.

A player can choose between one of two basic options in any interaction: 1) act as an individual; 2) cooperate with the other player. Acting as an individual may involve competing, showing a willingness to compete, or simply not cooperating and instead doing your own thing. It represents some form of rebellion against the other player's interests. The act of cooperation may involve flight from competition, showing an unwillingness to compete, or actively working with others. It's not the details that matter but the overall attitude toward rival players.

This book offers a simple and accessible introduction to the basic ideas and applications of game theory. First the basic principles of game theory are presented, from a game matrix and prisoners' dilemma, to a dominant strategy, Pareto efficiency and the Nash equilibrium. From there asymmetric payoffs and repeated games are explored, along with some of the general tactics and tricks that a player might use to improve their payoffs and their possible outcomes, based largely around the idea of asymmetric information. Mixed strategy and zero-sum games are also examined, with a look at well known games such as hawk-dove, chicken, and interactions between buyers and sellers.

The examples are made clear with over 50 images of the games throughout the book, and characters such as Jack and Jill, friend and stranger, and more are used to keep things interesting and bring the situations to life. The final section looks into some criticisms of the relevance of game theory, and addresses issues of altruism, hatred, and games that change in nature over time.

A Game Matrix

If life itself is a never-ending game then there can be no escape from acting as a player. That makes a focus on long-run outcomes that require investment pointless, as other players can ruin your plans at any moment. The only smart move must be to focus on the short-run process of playing the game, and always keeping your own interests in mind in any interaction.

A game matrix can be used to show the options available in an interaction, the range of potential payoffs on offer for those involved (the players), and the expected outcome. While the matrix is given for an interaction between two players it is applicable for up to any number, and essentially represents what to expect when there are divergent interests. The grid used shows the four possible outcomes that could arise in any interaction, and highlights the expected outcome and the resulting payoffs for players. The four potential outcomes in the game matrix are: both players cooperate; player one cooperates but player two rebels; player two cooperates but player one rebels; and neither player cooperates but instead rebel and seek to go their own way. This is a game matrix without specific players or their payoffs, to show the basic form of the model.

	Player 2 cooperates	Player 2 rebels
Player 1 cooperates	*(P2 payoff)* *(P1 payoff)*	*(P2 payoff)* *(P1 payoff)*
Player 1 rebels	*(P2 payoff)* *(P1 payoff)*	*(P2 payoff)* *(P1 payoff)*

There is a set of two payoff outcomes linked with all four of the possible options, and the lower left P1 payoff is the one for player one, while the upper right P2 payoff is always for player two. For example, in the situation where both player one and player two cooperate, the set of relevant payoffs would be the pair in the top left of the four pairs. And if both player one rebels and player two rebels, the set of applicable payoffs would be the one in the bottom right of the four pairs of payoffs.

The following game matrix adds numbers to the matrix to show what a completed game would look like. This is just an example matrix and the result will not be analysed or solved, but it shows the general idea of what to expect. It's assumed that players act in a way they think will give them the highest payoff, and that they have information as to the payoffs they receive from each possible outcome.

	Player 2 cooperates	Player 2 rebels
Player 1 cooperates	3 3	4 1
Player 1 rebels	1 4	2 2

The payoffs on offer here are 3, 1, 4 and 2. In ranked order from highest to lowest it's: 4, 3, 2, 1. Therefore players are expected to chase outcomes that give them a 4 payoff if available, a 3 as a second choice, their last resort is a 2 payoff, and a 1 is what they'll want to avoid as the lowest payoff. With this information some predictions could be made as to the result, but that's not for this section. The next section will use the same format as this game and apply it to a real life situation, following the players' payoffs to find the outcome and understand the applications of game theory.

The Prisoners' Dilemma

The prisoner's dilemma is perhaps the most famous example used in game theory, and sees individual incentives push players to betray the other when mutual cooperation would give both a far better result.

Two guilty suspects are being held and questioned by police, but they've agreed to keep quiet to reduce their chances of being found guilty and sent to prison. If they stick to this strategy (mutual cooperation) then the police only have the evidence to charge them both with a lesser crime, and they will serve a minimal sentence. But if one betrays the other (rebels) and places all the blame on them then he'll walk away free as an innocent man, while the other who holds to his word to keep quiet (cooperates) has the responsibility of two guilty men forced on him and suffers an increased sentence. If the two suspects both confess and betray (mutual rebellion) the other then they will serve the normal sentence for their crimes.

Prisoner one is known as Bubba and prisoner two as Tyrone, and numbers in the matrix below represent the hypothetical payoffs for each player for each course of action. Bubba has the rows across and Tyrone the columns down, with the payoffs for each outcome where the choice options meet. Bubba has the payoff numbers on the left of each pair, and Tyrone's are on the right.

	Tyrone keeps quiet	Tyrone betrays
Bubba keeps quiet	-1 / -1	0 / -10
Bubba betrays	-10 / 0	-5 / -5

This may appear confusing, as there are now negative numbers and a zero in the grid, in place of the positive numbers from earlier. But the principle is the same as in the last game matrix, and players seek the highest payoffs that they can get. In this game there are only four possible payoff numbers on offer to a player: -1, -10, 0 and -5. Ranking them in order from largest (what players want and pursue) to smallest gives: 0, -1, -5, and -10. Players will go for 0 as the preferred choice payoff if available, their next choice will be a -1 result, their last resort is a -5 payoff, and the -10 payoff is the one they're desperate to avoid as it's the worst possible outcome.

The matrix contains eight pieces of information that can be recorded.

If Bubba and Tyrone both cooperate and keep quiet:
Bubba's payoff = -1.
Tyrone's payoff = -1.

If Bubba cooperates and keeps quiet, but Tyrone betrays and implicates him:

Bubba's payoff = -10.

Tyrone's payoff = 0.

If Tyrone cooperates and keeps quiet, but Bubba betrays and implicates him:

Bubba's payoff = 0.

Tyrone's payoff = -10.

If Bubba and Tyrone both betray and implicate the other:

Bubba's payoff = -5.

Tyrone's payoff = -5.

If both prisoners rebel and betray the other, going against their cooperative agreement to keep quiet on what went down, there's enough evidence to find them guilty and they'll each be kept in prison for five years. This is the outcome of -5, -5 payoffs in the bottom right of the grid.

Prisoner one Bubba betraying prisoner two Tyrone and putting all the blame on him sees Bubba walk free, a payoff of 0 or no time in prison, and the hapless Tyrone who naively kept quiet serves a ten year sentence as he is punished for the crimes of two men, with a -10 payoff in the matrix. This is the 0, -10 outcome in the bottom left of the grid.

If Bubba keeps quiet and gets betrayed then it's the backstabbing Tyrone who walks free, while Bubba serves two men's five years sentences for ten years total, in the -10, 0 payoffs at the top right of the grid.

And if both keep quiet then the police can only charge them with a lesser offense and they both serve one year in jail, which is the -1, -1 outcome in the top left of the grid.

Looking at things from Bubba's point of view can show what to expect, and he will think about what the other suspect being questioned may do, as that will affect his own payoff. The game matrix below takes a small part of the larger matrix above, to see first what Bubba should do if Tyrone was to keep quiet as agreed. This section of the game is examined from the first person point of view of prisoner Bubba.

Tyrone keeps quiet

I keep quiet   -1

I betray   0

If Tyrone keeps quiet and cooperates then it's best for Bubba to rebel and betray him by incriminating the poor man. That earns him an individual payoff of 0, and no time at all in jail as he walks free, compared with a payoff of -1 and one year in jail if he sticks to his word and says nothing of their crimes. Next we'll look at what Bubba should do if Tyrone betrays him, and this again simply involves taking a small part of relevant information from the larger matrix above.

	Tyrone betrays
I keep quiet	-10
I betray	-5

In a situation where Tyrone betrays him Bubba knows that it's best for him to respond in kind. That would see him get a payoff of -5 and five year sentence, no worse off than he would be had he faced the police alone. But if he foolishly keeps quiet as prisoner Tyrone convinces the police that he is evil he'd get a payoff of -10, and serve ten long years in jail. Whatever Tyrone does the best response for Bubba is to rebel and betray him.

The strategy to rebel strictly dominates the strategy to cooperate for player one here; it provides a greater payoff for him irrespective of the actions of the other player. When one strategy strictly dominates another it becomes a dominant strategy for a player, and it's what we would expect them to do all of the time. The other type of dominance is weak dominance. When one strategy weakly dominates another it provides no worse a payoff, and can provide a greater payoff depending on the behaviour of the other player. In mathematical terms, strictly dominates means it is greater than ($>$), while weakly dominates means greater than or equal to ($\geq$).

Next we can look at the part of the complete game matrix that shows Tyrone's payoffs, and first what he should do if Bubba was to cooperate and keep quiet as previously agreed. As before this is how Tyrone would see it in the first person.

	I keep quiet	I betray
Bubba keeps quiet	-1	0

If Bubba keeps quiet then Tyrone should do the opposite and betray him, which gives a payoff of 0 and means no time at all in jail. This is better for him than keeping quiet in unison, as that gives a payoff of -1 and a year in jail.

	I keep quiet	I betray
Bubba betrays	-10	-5

If prisoner Bubba betrays him then it's best for Tyrone to do the same in return, which earns him a payoff of -5 and five years in jail, half the ten years in jail and -10 payoff he'd get for foolishly keeping quiet as his fellow suspect does the opposite. Whatever Bubba does it's best for Tyrone to betray and incriminate him during the police questioning, and that's his dominant strategy in all circumstances.

Putting the dominant strategies of both prisoners together gives the matrix below, with the expected outcome of mutual betrayal where both men earn payoffs of -5 and five years in jail each.

	Tyrone keeps quiet	Tyrone betrays
Bubba keeps quiet	-1 -1	0 -10
Bubba betrays	-10 0	<u>-5</u> <u>-5</u>

This result is known as the Nash equilibrium outcome, where neither player can improve their payoff without the other player changing their own too, and it is the stable long-run outcome. More specifically this is a pure strategy, sub-game perfect Nash equilibrium, as each player only ever has one strategy (to rebel) and every small part of the long-run game would follow this same pattern. This gives mutual rebellion and payoffs of -5 or five years in jail each.

Yet if both had stuck to the agreement to keep quiet then there wouldn't be enough evidence to prove their worst crimes. They would get only a -1 payoff each and one year in jail instead of the four more years suggested by the game. That would be the Pareto efficient outcome here, where no player would be better off without making another worse off. But individual incentives cause both men to ruin themselves in a 'rush to the bottom' scenario, in theory at least.

Another factor that makes mutual betrayal more likely is that this is a quite unique game, where a betrayal cannot be withdrawn and gives characteristics of a one-off game, while keeping quiet can be changed at any time the men are in custody. This may make the prisoners feel that keeping quiet is only a temporary strategy by the other prisoner, until he finds the best moment to stick the knife in, and they may want to do it themselves first to avoid being a victim.

In this case the perverse incentives explained by game theory are actually in the best interests of wider society. They see two guilty men both face full responsibility for their terrible crimes, instead of one walking free or the two of them only serving jail time for their additional minor offenses. But this will not always be the case and game theory is neither good nor bad, it simply shows what can happen as different players search for a better payoff.

The specific numbers in the game matrix above are not important to the result of mutual rebellion against the cooperative outcome, and all that matters is the overall ranking of the different results. Players will miss out on a mutually superior outcome due to individual incentives if the ranking of payoffs for a player are (from best to worst): 1) to rebel from a cooperating player; 2) mutual cooperation; 3) mutual rebellion; 4) to cooperate and then have the other rebel on you.

The Free Rider Problem

As has been shown above two players may rebel and betray the other when a better outcome could be achieved with cooperation. One of the most damaging examples of this occurs with the problem of free riders. This is a situation that can play out everywhere, from close relationships to passing acquaintances, and from the workplace to the wider environment. Readers may wonder why they sometimes end up in situations where despite intending to work hard and contribute, they find themselves holding back and acting as a freeloader. It's essentially a defensive manoeuvre, as the rest of this section explains.

In the earlier matrices the two options for a person (player) in an interaction (game) were to cooperate or rebel. In this section that largely remains the same but cooperate will be referred to as 'contribute' while rebel is now 'parasite' instead. An example game matrix showing the free rider problem is given below, using the same basic model as the prisoners' dilemma seen before, except this time payoffs are not symmetrical. The two players in this game are Jack and Jill, who went up a hill to fetch a pail of water. But now there is a question over who's going to be doing all the work to collect the water and then bring it back down the hill again.

	Jill contributes	Jill parasites
Jack contributes	3 2	4 -3
Jack parasites	-2 3	0 1

A player can either choose to put effort in and contribute to the interaction or to parasite and make no attempt to offer anything, instead focusing on taking what others have put in. If both players refuse to give anything at all and instead only try to parasite, then the outcome is that in the bottom right of the grid, with payoffs of 0 for Jill and 1 for Jack. The logic behind these values is that Jack will fare better on his own than Jill will. If both contribute then each player will receive what they have created, and part of the synergy gain that their interaction made. Payoffs will be 2 for Jack and 3 for Jill, as Jack is assumed to be the more efficient worker here, and any synergy gain will benefit her more than him relative to their contribution.

When one player contributes and the other parasites there is a disparity in the payoffs, with the parasite player able to enjoy the benefits created when someone else makes a contribution, without having to exert any effort of their own. If Jack parasites he gets a payoff of 3, but Jill will get 4 for doing the same thing as Jack produces the greater output of the two and there's more output to parasite. Meanwhile the poor player who contributed only to see others take it from them suffers a loss, and is filled with bitterness over a freeloader taking what they worked hard to produce. If Jill suffers this then she gets a negative payoff of -2, while Jack suffers ever worse as he had more to lose, and suffers a -3 payoff.

The game matrix below predicts what Jack will do, looking at the choice to contribute or parasite from his point of view.

	Jill contributes
I contribute	2
I parasite	3

As Jack sees it, if Jill contributes something to the interaction then it's in his own best interest to parasite, rather than join her in making a useful contribution. The parasite plan would earn him a payoff of 3, which is more than the payoff of 2 he gets for making an effort, as he would gain something without having to use up any of his own energy.

	Jill parasites
I contribute	-3
I parasite	1

In the alternative scenario where Jill gives nothing to the interaction and chooses to act as a parasite, Jack knows that the only smart move is to also parasite. If he does that then he can still produce a little for himself and get the payoff of 1. Yet if he was foolish enough to try to contribute while Jill is determined to do nothing more than parasite, then he'd end up worse than before he started, with a payoff of -3. No matter what Jill does it's best for Jack to parasite, and that is his dominant strategy.

Next we can examine the payoffs for Jill, from her point of view.

	I contribute	I parasite
Jack contributes	3	4

If Jack was to contribute then Jill would be best off by choosing to parasite. That earns her a payoff of 4 in place of the 3 payoff she'd get by copying Jack's effort to contribute, as she'd get resources without having to put in any effort of her own.

	I contribute	I parasite
Jack parasites	-2	0

If Jack was to parasite then Jill should copy his behaviour. That earns her a payoff of 0, and although it isn't a gain it's better than the -2 payoff that comes when she contributes as Jack resolves to do the opposite. Like Jack, Jill's dominant strategy is to parasite in all circumstances.

With this information we can predict the outcome and Nash equilibrium of the game, and it is underlined in the bottom right of the grid below.

	Jill contributes	Jill parasites
Jack contributes	3 2	4 -3
Jack parasites	-2 3	0 1

Perverse individual incentives here once again push both players to avoid making contributions to an interaction, and instead they are likely to try to parasite. The result is a very small payoff for Jack alone, and little is created. Jill is the biggest loser here as her end result is a payoff of 0. That's three below what it could have been, and Jack gets a payoff of 1, only losing a payoff of one unit compared to the two on offer from cooperation. But even though Jill has a lot to lose her individual incentives will still push her to try and parasite rather than contribute, and the fact that she gets the better deal from cooperation is irrelevant to her chasing a bigger and better payoff. There's no place for gratitude in game theory.

But just like the game earlier it would have been better for both if they had contributed, and asymmetric payoffs do not necessarily change this outcome of the game. The complete game matrix above shows that although Jack and Jill may not possess symmetrical payoffs, they still follow the payoff ranking of the prisoners' dilemma. The only difference between the two players is that Jack loses one less from his payoff than Jill when the other player parasites, while Jack gains one less for each outcome than Jill when the other player contributes, as he contributes more output.

The asymmetric payoffs of the type shown above are relevant to the world of work, and some workers feel resentment that others gain far greater rewards for contributing than they do. They may see their position of power, as Jack has above, and say to their boss 'you need me more than I need you, and I demand a better payoff' (i.e. more money). But individual incentives still suggest that the worker won't get what he feels he deserves, and his effort may fall as a result.

If two men were to work together and contribute effort to fix a road then both may benefit by being paid a decent wage. But the more productive of the two may become resentful at not being rewarded for his contribution, and instead seek to shirk his duty and try to parasite from the other, as that is the only way to be sure to prevent it happening to them. Then the other has reason to respond in kind. No man wants to work while his fellow man sits back and takes all of the credit, as that essentially makes the former the slave of the latter. And if it's a choice between being a slave or a parasite, then many may understandably choose the second option as the lesser of two evils.

Not only is this outcome worse for the two players involved that mutual contribution would have given them, it's also worse for society. When players contribute to an interaction there will be positive externalities that are absorbed by the wider community, and anything created by either man that isn't used by one of them is an externality. But when neither contributes at all there are none.

The problem here is that this is essentially an endlessly repeated game, where a man could change his choice from contribute to parasite, or switch the other way, at any time. That means that two men may work together and both contribute, but when one turns his back to do more the other can slack off and start to parasite instead. This can lead a player to want to see the other begin work before he does, which means that work is reluctant and slow, with the potential levels of output and outcome never being realized.

Strategic Codependence

Game theory suggests that although it may be in players' mutual interests to cooperate, their individual incentives can prevent this from happening as successfully getting one over the other player gives a higher payoff. This has implications for all interactions, from those between national governments at the highest level of global politics, to those among citizens trying to coexist in a civilized society, and everywhere else in between.

In a prisoners' dilemma game the cooperation (keep quiet) option is given as a repeated choice that can be changed later, while the rebel option (betray) is a one-off that can't be revoked but can be used as a well-timed tactical strike on a rival. These characteristics see the Nash equilibrium of mutual rebellion. In a free rider game the cooperation (contribute) and rebel (parasite) options are both repeated choices that can be changed at any time, and this allows a player to choose to parasite soon after the other contributes. The nature of rebelling is different here, but the Nash equilibrium is once again for both players to rebel.

One thing that hasn't been looked at yet is a game where the nature of cooperation is different, and if other factors don't affect the Nash equilibrium long-run outcome then this area might. Perhaps a game would have a different result if cooperation wasn't a repeated option that could be abandoned later, but a one-off commitment that couldn't be changed.

The natural solution to the problem of individual incentives preventing a mutually beneficial collective goal is to take away the power to act on those incentives. That may involve a higher body that intervenes when individual countries threaten to step out of line and attack each other, or when citizens abuse those who have less power than they do. In both cases the strategy is to force codependence on players, to remove their ability to rebel from the mutually beneficial outcome of cooperation.

The standard prisoners' dilemma game matrix can offer a reminder of the payoffs and long-run outcome before strategic codependence is forced on players. The two players here are Ken Right and Bill Goodman, who are two self-righteous men holding authority at the decision making level of two national governments, and they aim to do the best for their countries in the international arena. Both players will naturally rebel against each other to follow their own interests as a dominant strategy, as they believe they know what's best for their countries, which would lose global power and influence if they don't stand up for them internationally. This gives the sub-optimum 1, 1 payoff underlined.

	Bill Goodman cooperates	Bill Goodman rebels
Ken Right cooperates	2, 2	-1, 3
Ken Right rebels	3, -1	<u>1</u>, <u>1</u>

Next the world the game takes place in will be changed while the payoffs remain the same. The assumption now is that national governments are forced to turn to the United Nations before selecting an interaction strategy with other countries. The amended game matrix now has only two possible outcomes instead of the four seen in earlier games.

	Bill Goodman cooperates	Bill Goodman rebels
Ken Right cooperates	<u>2</u>, <u>2</u>	
Ken Right rebels		1, 1

There are now only two options for Ken Right and Bill Goodman; it's either simultaneous cooperation or simultaneous betrayal. Either their countries cooperate explicitly on certain issues or they are open enemies. That leaves cooperation as the new best choice and it's now the dominant strategy for both players. This makes the underlined payoffs of 2, 2 the new Nash equilibrium of the game, and both players achieve a far better outcome than before.

Mutual cooperation could involve the UN members watching out for a basic international code of conduct, or the countries of Ken Right and Bill Goodman working together on an agreed course of action. Mutual betrayal may mean that a country goes against the collective's values, who then respond in kind with some form of punishment.

The only question is whether this is genuinely possible, and if the outcome where one player/country cooperates in an interaction as the other rebels could really be removed by a collective authority. For that to be the case the rule would have to be that you either cooperate to be welcomed as part of the community and receive the benefits, or you betray the group interest and lose membership and all of the associated privileges. On the face of it this seems quite possible, as anyone who attacks another nation without justifying it to the UN is likely to face their sanctions, just as a vigilante attacking anyone in the neighbourhood who offends them will be targeted by an effective neighbourhood watch group.

In reality however things are not quite that simple. Although there may ultimately be agreement at the community level or at the UN, there is still the issue of who decided the agenda that was agreed upon, as the details of the collective proposal will be disputed. Ken Right and Bill Goodman will want to betray each other and anyone else present, and have everyone cooperate with their worldview. They'll try their hardest to resist losing influence in the community arena. And even with an agreement in place on UN or on neighbourhood watch action, the two men and those they have authority over will still naturally try to avoid contributing their resources, and instead try to parasite protections while others do the work. The payoffs linked to deciding the group agenda, or giving resources to enforce it, are represented in the game matrix below.

	Bill Goodman cooperates	Bill Goodman rebels
Ken Right cooperates	2 2	3 -1
Ken Right rebels	-1 3	$\underline{1}$ $\underline{1}$

This is the exact same situation as before and forced codependence has changed nothing. With all four outcomes back in the matrix the dominant strategy is once again for both players is to betray the other, and the Nash equilibrium is mutual betrayal and lower payoffs of 1, 1 than the 2, 2 payoffs on offer with cooperation. Even with a higher authority enforcing rules from above players will still rebel during the process of agreeing rules, and also in the process of enforcing them. It seems there's no getting away from the game no matter what the type of interaction, and there's no easy way to simply remove the undesirable option of one player betraying another.

Repeated Games and Tit for Tat

It may be difficult to stop a player rebelling from the cooperative option for personal gain, making things worse for everyone in the process. But while in the classic one-off prisoners' dilemma game there is no way around this, in an iterated prisoners' dilemma this can potentially be overcome. In repeated games the other player has the choice to respond in kind to punish the rebelling player. In areas based around prolonged and not one-off interactions the rebelling player has to remain around the player he has just betrayed, who can then ruin his plans of achieving the game's maximum payoff.

A tit for tat strategy is where a player openly states his intention to do whatever the other person did in the previous round of the game. Consider two players having repeated interactions in a relationship; player one is Man and player two is Woman. If Man cooperated, then Woman will respond in kind the next chance she gets, but if she sees a move to rebel then she will follow suit and do that instead. And if Woman resolves to adopt such a zero tolerance strategy, then it makes sense that Man would also follow the tit for tat strategy, to avoid the negative payoff associated with cooperating with a rebelling player.

Players will think ahead to see if rebelling is more rewarding than cooperation when the other player follows tit for tat, and they'll use what is known as forward induction to predict payoffs in future rounds. The two players, Man and Woman, will use the same game matrix as in the previous section.

	Woman cooperates	Woman rebels
Man cooperates	2, 2	-1, 3
Man rebels	3, -1	1, 1

The payoffs on offer to a player are as follows: rebel as rival cooperates (RC) = 3; both cooperate (CC) = 2; both rebel (RR) = 1; cooperate as rival rebels (CR) = -1. Although payoffs in other games for the four possible game outcomes may differ from those here, the key point is that CR gives the worst result, with RR a little better, CC a still better payoff, and RC offers the best result.

If Man and Woman start the first round of the repeated game cooperating, and follow a tit for tat strategy thereafter for additional rounds, then the following results would be expected after three rounds of interaction:

Round 1 = Man cooperates (CC payoff 2), Woman cooperates (CC payoff 2);

Round 2 = Man cooperates (CC payoff 2), Woman cooperates (CC payoff 2);

Round 3 = Man cooperates (CC payoff 2), Woman cooperates (CC payoff 2).

Total payoffs after three rounds of interaction are 6 for both Man and Woman. Three rounds can also be examined if Woman chooses to rebel once in round two, before accepting responsibility for her mistake to cooperate the next round and take her punishment of tit for tat. After that both players will return to cooperating in the unseen round four and then follow tit for tat, exactly as above. The results for the first three rounds are:

Round 1 = Man cooperates (CC payoff 2), Woman cooperates (CC payoff 2);

Round 2 = Man cooperates (CR payoff -1), Woman rebels (RC payoff 3);

Round 3 = Man rebels (RC payoff 3), Woman cooperates (CR payoff -1).

Total payoffs after three rounds of interaction are 4 for both Man and Woman. Having just one further round after a player rebels is enough to see that person suffer a worse outcome than if they had cooperated throughout, as long as the player who rebels takes responsibility for their actions.

With this in mind the rebelling player may avoid taking responsibility, and instead of cooperating to let the other player 'screw them over in return' they may simply return to their tit for tat strategy and pretend they'd done nothing wrong:

Round 1 = Man cooperates (CC payoff 2), Woman cooperates (CC payoff 2);
Round 2 = Man cooperates (CR payoff -1), Woman rebels (RC payoff 3);
Round 3 = Man rebels (RC payoff 3), Woman cooperates (CR payoff -1).

Total payoffs after three rounds of interaction are 4 for both Man and Woman, once again giving a worse payoff for the rebelling Woman than the continued cooperative outcome. The result is exactly the same as if Woman tried to make up for her behaviour, as the return to tit for tat in round three sees Woman mimic the cooperation of Man in the round before. Woman will therefore know that pretending to have done nothing wrong and continuing as before doesn't earn her a better outcome, and nor does apologizing to Man and trying to make amends. With this in mind Woman may decide to rebel and commit to that policy. It is worth looking at the payoffs where Woman rebels from round two onwards while Man follows tit for tat, to be sure there is no incentive for that:

Round 1 = Man cooperates (CC payoff 2), Woman cooperates (CC payoff 2);

Round 2 = Man cooperates (CR payoff -1), Woman rebels (RC payoff 3);

Round 3 = Man rebels (RR payoff 1), Woman rebels (RR payoff 1).

Total payoffs after five rounds each are 2 for Man and 6 for Woman. Woman gets a higher total payoff than Man but that's not important, and what matters is that the overall payoff for her is no higher than the 6 on offer if she cooperates in all rounds. That is true in this extended game with additional rounds of interaction, where player two Woman has to suffer at least one further round of player one Man's response. This suggests that the tit for tat strategy is an effective one, and it has removed the temptation for the other player to rebel.

While a tit for tat strategy could act as an incentive for players to cooperate and not rebel, once a player rebels just once there is no returning to mutual cooperation under the system, and all future gains are gone forever. If they were to do it by accident or in confusion then a resolution would be unlikely.

There are also two further problems with the system. First of all, tit for tat may not work properly if the payoff for rebelling on a cooperating player (RC) is significantly better than the mutually cooperative (CC) outcome, or if the payoff from mutual rebellion (RR) is not much worse.

In the example above RC is 3, CC is 2, and RR is 1. That means that the player who rebels will get RC = 3 in the round they do that, and RR = 1 the next round if they do the same again (as they should) and the other player responds in kind. 3 and 1 make 4, just as two rounds of mutual cooperation CC = 2 give 4, and the player who rebels is not rewarded. But if (RC + RR) exceeds (CC + CC) then a tit for tat strategy will not work, and it's still best to rebel even after tit for tat punishment.

The biggest problem however is that the tit for tat strategy only works if there are going to be future rounds of interaction. Without that then there can be no punishing a player who rebels, and there is likely to be more if it.

Backward Induction

The duration of an interaction is very important, and a one-off game may be destined to see a player rebel as there's no fear of a backlash from the other player, but mutual cooperation could be incentivized in a repeated interaction with a tit for tat strategy. However, even in a repeated game a player may be able to avoid the problems caused by a tit for tat system and achieve a superior payoff, if he can use 'backwards induction' to decide his best move in every round. This is where a player looks to the future to decide their best move in the final round of the game, and then works backwards to find the most rewarding move in the round before, and so on to the start of the game.

For the purposes of this section the two players are called Nostradamus and Cassandra, and in the final round F of a game the interaction is essentially a one-off. Individual incentives push both players to rebel as the game follows the same form as seen earlier, and with no further rounds of interaction there is no need to fear a tit for tat reprisal. There is also no point trying to force a cooperative outcome, as above with strategic codependence, as it isn't worth the effort when there will be no further interactions. With the dominant strategy to rebel as shown in the following game matrix, there is a

Nash equilibrium result of mutual rebellion and 2, 2 payoffs in this example.

	Cassandra cooperates	Cassandra rebels
Nostradamus cooperates	4, 4	-4, 6
Nostradamus rebels	6, -4	<u>2</u>, <u>2</u>

Because both players know that the final round F will inevitably see individual rebellion, the interaction one round earlier, F minus one or (F-1) for short, is essentially the last round. This is where they'll need to make their last decision and it becomes the final round in practice.

Round F = Nostradamus rebels, Cassandra rebels;
Round (F-1) = Nostradamus ?? , Cassandra ??

As both players make their decision of what to do in round (F-1) they will think over their incentives and the consequences of their choice. Their incentives are the same as always: to rebel. But they may worry about the opportunity cost of this choice for the future, and whether it would a better idea to sacrifice short-run interests for long-run interests with cooperation, instead of inviting the other player to punish their rebellion with a tit for tat rebellion. Yet a look ahead to the next round F shows that whatever they do the other player will be certain to rebel, effectively punishing any attempt to cooperate. With the other player following that path there's no reason to hold back, which will see both players acting on individual incentives to rebel in round (F-1).

If final round F and the (F-1) round before it will always see mutual rebellion then the interaction before either, round F minus two or (F-2) for short, becomes the last round requiring a decision from the two players.

Round F = Nostradamus rebels, Cassandra rebels;
Round (F-1) = Nostradamus rebels, Cassandra rebels;
Round (F-2) = Nostradamus ?? , Cassandra ??

But the same logic applies here as it did to round (F-1), and knowing that all future rounds will see mutual rebellion they have no reason to not act on their incentives to rebel now. This pattern will repeat right to the start of the game, and players can be expected to rebel in every single round. This is a big concern, and it suggests that rebellion is not a brief problem that can be resolved for superior payoffs, but something that will remain an issue throughout the life of the game. Yet there is a way around this, and it requires a closer look at backwards induction.

Readers are likely to have used backwards induction in their own lives, and everyone does it when they want to figure out how to achieve a specific outcome. First a person targets a goal and then they think about the steps that lead there. For example, someone who travels to their workplace may go through the process as follows.

Arrive at the workplace door
(FOLLOWS)
Leave car and walk to entrance
(FOLLOWS)
Drive car along known route
(FOLLOWS)
Leave home and get in car

In this example the targeted goal and long-run outcome is an individual one, and there are no interactions with others to throw the plan off track. All it takes is to know the desired goal and then to work backwards in simple steps. Readers will have done this countless times and it feels easy and becomes second nature. But if backwards induction is causing players to rebel in every round of the game, then it should be equally easy to throw a spanner in the works and prevent it from happening. If a player is prevented from knowing his desired goal then he can't apply backwards induction and can't rebel.

A player will always rebel in the last round, and then work backwards to do the same in all previous rounds as a result, but he can't be sure to rebel if he doesn't know when the end of the game will be. The idea is shown below.

Round ?? = Nostradamus ?? , Cassandra ??
Round (??-1) = Nostradamus ?? , Cassandra = ??
Round (??-2) = Nostradamus ?? , Cassandra = ??

In this example 'Round ??' is an unknown round number, where the question mark may represent the final interaction or an early one, while (?? -1) and (??-2) are the two preceding stages. If 'Round ??' is the last round then the smart move is to rebel, which makes 'Round (??-1)' the acting last round where players should rebel, and so on. But if 'Round ??' was an early round, and the other player cooperated and employed a tit for tat strategy to deter rebellion behaviour, then it may not be best to rebel at all.

The following game matrix gives a reminder of the payoffs from a one-off game in this case, and using these payoffs the overall return from five rounds of interaction can be calculated for players, assuming that 'Round ??' is an early round, and there are at least five rounds of the game ahead.

	Cassandra cooperates	Cassandra rebels
Nostradamus cooperates	4 4	6 -4
Nostradamus rebels	-4 6	2 2

Player one Nostradamus is unaware of the end of the game and therefore cooperates throughout, covering himself against a tit for tat response that would reduce his payoffs (2 in place of 4) in case it's an early round of the game with many rounds ahead. He also employs tit for tat to punish rebelling behaviour from the other player to stop them from 'cheating'. But here Cassandra knows that there are exactly five rounds left and she plans to rebel in the last one as is in her best interest, and she uses backward induction to rebel in the four before it too as a result.

Round 1 = Nostradamus cooperates (CC payoff -4), Cassandra rebels (RC payoff 6);

Round 2 = Nostradamus rebels (RR payoff 2), Cassandra rebels (RR payoff 2);

Round 3 = Nostradamus rebels (RR payoff 2), Cassandra rebels (RR payoff 2);

Round 4 = Nostradamus rebels (RR payoff 2), Cassandra rebels (RR payoff 2);

Round 5 = Nostradamus rebels (RR payoff 2), Cassandra rebels (RR payoff 2).

After five rounds of interaction Nostradamus has a payoff of 4, and Cassandra gets a payoff of 14. But if Cassandra was also unaware of the timing of the end of the game, and didn't know if it would end at all, then she may have acted differently. She would be unable to use backward induction without a point to work back from, and all that would be left would be forward induction as seen in the last section. And forward induction would tell her that she would lose superior cooperative gains (4 each round instead of 2) if she rebelled too early, and even if she plans to rebel in the future it's best to delay it as long as possible. The example below shows her payoffs if she left the rebel move until the last round of the five.

Round 1 = Nostradamus cooperates (CC payoff 4), Cassandra cooperates (CC payoff 4);
Round 2 = Nostradamus cooperates (CC payoff 4), Cassandra cooperates (CC payoff 4);
Round 3 = Nostradamus cooperates (CC payoff 4), Cassandra cooperates (CC payoff 4);
Round 4 = Nostradamus cooperates (CC payoff 4), Cassandra cooperates (CC payoff 4);
Round 5 = Nostradamus cooperates (CR payoff -4), Cassandra rebels (RC payoff 6).

After five rounds here Nostradamus has a payoff of 12, and Cassandra has a payoff of 22, far higher than before. By cooperating through the first four rounds Cassandra doesn't activate Nostradamus' tit for tat strategy, but the rebel in the final round of the five sees her get the highest possible payoff.

If a player doesn't know where the current round of interaction figures in the overall lifespan of the game they also can't know when the end of the game will occur, and that challenges backwards induction. They will be wary of choosing to automatically rebel as that puts potentially higher payoffs at risk as noted above, and will be left unsure how to proceed, and unable to use backwards induction to rebel in every round.

For a player to be unaware of the current round's position in the span of the game they must be unaware of the nature of the interaction and unsure when it begins or ends. The less information they have the better, and two strangers fit the bill perfectly. Many people will treat total strangers far better than those closest to them for a simple reason; they are wary of making a misstep as they don't know the nature of the game or the duration of the interaction. The less information players have about the other the less likely they are to rebel, as they can't be sure of the payoffs on offer. This suggests that payoffs are not a fixed factor but something that can be hidden or manipulated, and that could be exploited by smart players.

Cartels

Game theory suggests it's very difficult to curb individual incentives to rebel and screw over the other player, ruining the mutually beneficial cooperative outcome in an attempt to gain higher individual payoffs. But perhaps all of the focus on trying to manage the game is targeting the wrong issue, and a better place to start may be trying to actually change the payoffs that players get from the game.

The reason a player rebels is because the individual payoff associated with that strategy is greater than that from cooperation, and to ensure the mutually beneficial outcome of combined cooperation the challenge is to change this. This may come from lowering the high payoff coming from rebellion, or it might be possible to raise the payoff linked to cooperation. When players see that rebellion doesn't pay in the long-run then they may be willing to make the effort to change the game, to prevent it happening in the future.

While a game is between two sides there are always external factors at play too. In the prisoners' dilemma game it was the police and jailors, but it can also be the general public. The game matrix below introduces two fictional competing big companies who sell the same products to customers, referred to as Talmart and Sesco. For the purposes of this game we could see them as selling the same branded products to the general public. The payoffs from this game relate directly to profits made from different pricing strategies, and players can either maintain their current price levels, or lower their prices to try and undercut their rival and get more business from customers.

	Sesco maintain prices	Sesco lower prices
Talmart maintain prices	4, 4	5, -2
Talmart lower prices	-2, 5	<u>2</u>, <u>2</u>

This is the basic prisoners' dilemma game, and as it stands at the start before intervention both players have the dominant strategy to lower prices to undercut their rival. This would earn Talmart and Sesco a 5 payoff each in place of the 4 on offer from cooperation. The logic being that the greater sales numbers won from the rival more than make up for lost income per unit. But when both do it sales don't increase, as there is only a certain level of demand for the products and the sellers would be sharing it, and there are lower profits as customers spend less for each unit. Yet this underlined 'price war' scenario is the result of individual incentives to compete with the other on price, with the Nash equilibrium outcome where each only earns a payoff of 2, less than the 4 on offer had they cooperated.

Yet if Talmart and Sesco could get together they may increase their gains by achieving the superior mutually cooperative payoffs. But they could only do this if their incentives pushed them that way, and they need to remove the temptation to rebel and compete. Raising the payoffs from cooperation directly may be beyond them as the only clear way would be to engage in a plan of price-fixing, and that's out as it's likely to attract the attention of the authorities. Even if it didn't the incentives would always be there to lower prices to secure a larger share of buyers' sales. The only realistic option is to reduce the payoff associated with lowering prices. This means that the two companies need to act like a type of cartel, despite their competitive position and lack of power over each other to enforce such a situation.

Lowering prices is always likely to gain sales and income of course, so the only way to reduce this payoff is to increase the cost associated with lowering prices. One option is an expensive marketing and branding campaign by both sellers that links their branded identical product to a specific memorable price, such as $9.99. In that case any price cut would require an expensive new campaign to replace the old one. Or alternatively the sellers could figure out the rough amount of sales they expect to make at such a price, and only have that many manufactured and ready for sale in stores or online. Any price cut in those circumstances would see lower prices without increased sales as the products wouldn't be available to sell, and any attempt to change this situation would of course also cost money. There would always be the risk that one company would do all of this while the other didn't, and would still have the incentive to lower prices, but the methods here are quite visible and transparent and the firms could observe their rival's investment or lack of it.

If these strategies were successful the game would look like the one below.

	Sesco maintain prices	Sesco lower prices
Talmart maintain prices	4 4	3 -2
Talmart lower prices	-2 3	0 0

The payoffs from lowering prices have been reduced by a payoff of 2 in all circumstances, due to the costs in doing so mentioned above. A one sided price reduction now has a payoff of 3 instead of the 5 earlier, and mutual price cuts now have a payoff of 0 instead of the 2 given before.

With the change to the matrix it's worth running through the game from the players' point of view, starting with seller one Talmart.

	Sesco maintain prices
We maintain prices	4
We lower prices	3

If seller two Sesco maintains prices then seller one Talmart knows that they should maintain prices too. That will give the Talmart Company a payoff of 4 that exceeds the 3 payoff linked to lowering prices.

	Sesco lower prices
We maintain prices	-2
We lower prices	0

And if Sesco lowers prices then Talmart should lower prices as well, to gain the better payoff of 0 compared to the payoff loss -2. Talmart's best response strategy is to do the same as Sesco, and with the game matrix symmetrical Sesco's strategy is to do the same as Talmart.

This gives a strange result, one not seen so far in this book. There is no dominant strategy for a player and no single Nash equilibrium outcome. Instead each player has a mixed strategy; where a player's best move depends on the action taken by the other player. This gives two possible Nash equilibria as underlined below.

	Sesco maintain prices	Sesco lower prices
Talmart maintain prices	<u>4</u> <u>4</u>	3 -2
Talmart lower prices	-2 3	<u>0</u> <u>0</u>

A player in this game will simply do whatever he thinks the other player will, and that creates two possible long-run outcomes. Either the two companies will end up both maintaining prices, to get payoffs of 4 each, or they'll both lower prices to get 0 payoffs each. Once they arrive at either point they may be stuck at that there, as a player can't leave that Nash equilibrium and get a better result without the actions of the other player.

Which outcome comes to pass will depend on the two players and the external environment. The starting point and default should be to maintain prices, with the 4 payoff on offer here exceeding all others. But external pressure to lower prices from a government intervening to protect consumers, customers themselves who won't buy at higher prices, or other sellers on the scene who didn't raise their own prices, may see an effort by players to lower prices and give the 0, 0 payoffs. And if Sesco for example was to suspect Talmart of preparing to lower prices, and that they had deliberately avoided making this a costly option despite their pretentions otherwise, then Sesco will lower their own prices to again result in these lower payoffs. All it would take for cooperation to be elusive and the game to be ruined is misinformation about the other player.

Asymmetric Information

When players in the game of human interaction have a dominant strategy it doesn't matter what the other player does, and there's no need to even think about it. But if there isn't a dominant strategy but a mixed strategy then it's in their best interests to find as much out about the other player as possible, and to try and determine what they're likely to do as that decides their own best course of action. This is where the problem of the unknown comes into play. In a world of incomplete information they're unlikely to know for sure what the other player will do, and the power of asymmetric information will come to the fore. Player one knows his incentives and what he will do, but player two doesn't, and the same applies in the opposite direction. This allows for one player to manipulate another, and to hide his true motives by concealing his true payoffs and incentives.

In the game that follows player one has taken on the neutral role of 'Stranger' while player two is putting on an act, and playing the more cooperative role of 'Friend' here. This is the same game matrix that has been seen before with the true payoffs for each player, and both have a dominant strategy to rebel.

	Friend cooperates	Friend rebels
Stranger cooperates	4, 4	5, -2
Stranger rebels	-2, 5	<u>2</u>, <u>2</u>

The two players known as Stranger and Friend both understand the individual incentives at play here, but are powerless to resist them and will end up with the 2, 2 payoffs from mutual rebellion.

However, it's possible that Friend could lie about his own incentives, to trick Stranger into cooperation. In a world of incomplete information Stranger can't know for sure what Friend's incentives are, and with a lack of information he may naturally assume that other players have the same incentives as he does, and that Friend would rebel no matter what. But if Friend could signal that he received a far higher payoff from cooperation, then that could change everything.

If Friend had the same payoffs and incentives as player one, as in the matrix above, then both players would rebel from the start.

Round 1 = Stranger rebels (RR payoff 2), Friend rebels (RR payoff 2);

Round 2 = Stranger rebels (RR payoff 2), Friend rebels (RR payoff 2);

Round 3 = Stranger rebels (RR payoff 2), Friend rebels (RR payoff 2).

Payoffs here are 6 each for both Stranger and Friend after three rounds of interaction.

But if Friend acted like he is needy and vulnerable on his own, and a 'nice guy' who loves to cooperate, and acts like he always follows a strict moral code that abhors rebellion and abandoning others, then the payoffs above could get far better for him. If the act was convincing enough then Stranger may wrongly believe that the game matrix had payoffs as below.

	Friend cooperates	Friend rebels
Stranger cooperates	6 4	0 -2
Stranger rebels	-2 5	2 2

In this 'false' game matrix that player one mistakenly sees as the real thing, his payoffs are exactly the same as before, but there has been a change in player two Friend's payoff values. Successfully cooperating now earns the convincing actor 'Friend' a payoff of 6, higher than the 4 previously as he is now thought to be a passionate team player. And rebelling on Stranger's cooperation earns Friend a neutral 0 payoff instead of 5 before, as it is thought to go against his principles to take advantage and screw another player over. The other two payoffs of Friend, where Stranger rebels, have not changed as mutual cooperation was never possible.

If Stranger wrongly believes this game matrix to be the real thing, then as far as he is concerned Friend's decision process is as shown below.

	I cooperate	I rebel
Stranger cooperates	6	0

It looks like Friend will cooperate if Stranger does himself, as that earns him a far higher payoff.

	I cooperate	I rebel
Stranger rebels	-2	2

If Stranger rebels then it looks like Friend will also rebel, just as before. It seems that instead of a dominant strategy to rebel it's now in Friend's best interests to copy Stranger's move in the interaction, and he has a mixed strategy dependent on Stranger's behaviour.

When Stranger is tricked by Friend's act and wrongly assumes that the payoffs above are real and not just imagined, then he will expect mutual cooperation to be possible. He'll believe that the only two options Friend is interested in are mutual cooperation or mutual rebellion, as these offer him the highest payoffs. And this codependence puts all the power in the hands of the other player, and it's all up to Stranger which outcome will occur.

Friend cooperates

I cooperate   4

I rebel   5

If Friend cooperates then it's in Stranger's best interests to do the opposite and rebel, for the higher payoff of 5 instead of 4.

And if Friend chooses to rebel then it's also in Stranger's best interest to rebel, to secure a payoff gain of 2 instead of a negative payoff of -2.

	Friend rebels
I cooperate	-2
I rebel	2

Putting all of the imagined payoffs together, it seems that both players will copy the strategy of the other, except when Friend cooperates then Stranger will rebel. Holding incorrect beliefs about Friend's payoffs, Stranger believes that three rounds of interaction will follow as below, where the other player 'Friend' will try to cooperate at first for a higher payoff, and then give up for mutual rebellion. He believes he will get a payoff of 9, and isn't concerned about Friend's payoff, only his strategy which can have an effect on his own payoffs.

Round 1 = Stranger rebels (RC payoff 5), Friend cooperates;

Round 2 = Stranger rebels (RR payoff 2), Friend rebels;

Round 3 = Stranger rebels (RR payoff 2), Friend rebels.

But if Stranger is forward thinking then he may try to overcome his short-run interests. If he doesn't then he believes Friend's tit for tat strategy to copy whatever he does will make him suffer reduced payoffs in future rounds. Still holding false beliefs about Friend's desire to cooperate, Stranger may do the same in an attempt to create synergy gains and secure a higher payoff. Stranger may believe that a payoff gain of 12 is possible over three rounds if he were to cooperate, as Friend should follow the example as his payoff incentives appear to suggest.

Round 1 = Stranger cooperates (CC payoff 4), Friend cooperates;
Round 2 = Stranger cooperates (CC payoff 4), Friend cooperates;
Round 3 = Stranger cooperates (CC payoff 4), Friend cooperates.

Player one 'Stranger' believes that the ball is in his court here, and that he can cooperate with Friend for a regular payoff of 4, and when the interaction is coming to an end (sometime after these three rounds) rebel for his maximum payoff of 5. But if Stranger does this he may get a shock in round one, as reality turns out to be different to his imagined outcome.

Round 1 = Stranger cooperates, Friend rebels.

This is not what Stranger had in mind, as he believes that Friend only gains a positive payoff with a dual strategy that copies his own, either mutual cooperation or mutual rebellion. Player two, the convincing actor known as 'Friend', can exploit Stranger's naivety by claiming that it wasn't his fault that he didn't cooperate, he simply thought that Stranger would rebel and copied that strategy. This argument is plausible to the misguided Stranger, who thinks that Friend is an honourable character who gains nothing from independent rebellion. And Stranger is likely to reaffirm his desire to cooperate and then give Friend another chance in round two.

Round 2 = Stranger cooperates, Friend rebels.

But the same thing happens yet again. Just as before Friend will blame the poor outcome entirely on Stranger not proving himself, and insist that all he wants is to cooperate and that it's down to Stranger to make it happen. Stranger may even fall for it a second time, unwilling to accept that his beliefs of the other player are completely wrong and that he has been tricked form the start, and played for a fool.

Round 3 = Stranger cooperates, Friend rebels.

And in round three Friend rebels yet again. Yet Stranger may still consider this a misunderstanding, and there can be no other explanation as he sees it, with his belief that Friend loves cooperation and hates to rebel. He will be sure that the problem is only temporary before the two of them get back on track, and then he can enjoy payoffs of 4 every round from mutual cooperation as planned, which would surely be worth all of this hassle in the short-run.

	Friend cooperates	Friend rebels
Stranger cooperates	6 4	0 -2

But Stranger has the payoff values all wrong, and Friend knows exactly what he's doing.

	Friend cooperates	Friend rebels
Stranger cooperates	4 4	5 -2

Friend's real payoffs are very different from those he's been trying to convince Stranger, and are in fact identical to Stranger's own. Every round for three separate stages he's been gaining the maximum possible payoff of 5, and forcing the lowest possible payoff of -2 on his rival. At some point Stranger will figure out he's been played by Friend and the game should revert to a mutual rebellion outcome, as the real payoffs suggest, and there's no way that Friend will be able to keep playing that role forever. Soon enough even the most trusting of players will notice the difference between the act 'Friend' puts up and his actual behaviour, but Stranger has already suffered as Friend gained. And it never would have been possible without the latter holding asymmetric information about himself that the other player didn't possess.

Marketing

The section above looked at the scenario where a player would misrepresent his own payoffs for cooperation or rebellion, to get the other player to act in a more desirable way. They may exaggerate their own gains from cooperation or downplay their payoffs from rebellion, all in an attempt to get the other player to cooperate so they could then rebel for their highest individual payoff.

But the most common method used to secure a higher payoff is to convince the other player that it's in their best interests to cooperate. The entire marketing and advertising industry is based around this idea, with the emphasis either on how much better your life will be if you cooperate with them and buy their product or service, or how your life will get worse in the future if you don't buy their product and instead do your own thing. And with incomplete and asymmetric information it's easy to fall for it.

The following game matrix may be the reality of buyer-seller interactions, before any marketing or advertising takes place. The buyer wants to rebel and pay less than the product's worth, while the seller wants to rebel and charge more than it's real value. This is the standard prisoners' dilemma where individual payoff incentives see the dominant strategy of each player to rebel, which gives an outcome of mutual rebellion in the bottom right of the grid. This is worse for each player than the payoffs if they had both chosen to cooperate and make the trade, and offers a neutral payoff of 0 each.

	Seller cooperates	Seller rebels
Buyer cooperates	2 2	3 -1
Buyer rebels	-1 3	<u>0</u> <u>0</u>

But a marketing or advertising campaign may insist that the seller's product is special and unique, offering new benefits for all those who come into contact with it. They may show actors who appear happy with their lives and without the incentives to cooperate with sellers, like the buyer in the matrix above, but who suddenly see unexpected great payoffs when they gave the seller a chance and finally cooperate to buy their product. A good marketing campaign or effective advertiser can convince buyers that game payoffs are different than they actually are in reality, and the imagined payoffs may suggest that cooperation with a seller is the smart move for a prospective buyer.

	Seller cooperates	Seller rebels
Buyer cooperates	<u>4</u> <u>4</u>	3 -1
Buyer rebels	-1 3	<u>0</u> <u>0</u>

In this amended game matrix that a buyer mistakenly believes true if he falls for the product-hyping marketing campaign, the payoffs from mutual cooperation have been changed from 2 each to 4 each, and it's now best for the buyer and seller to cooperate. That means the seller sells their product and the buyer buys it. But there are two underlined Nash equilibria here just like in the cartel game above, with both mutual cooperation and mutual rebellion acting as long-run Nash equilibrium outcomes, where neither player can singlehandedly improve their payoff.

With the 4, 4 payoff change a buyer will now copy the strategy of the seller to cooperate or rebel, as seen in some previous games. This means that for the mutually cooperative outcome to occur it's not enough to just convince the buyer that his own payoff will be better, and that buying the product is the only way to get the highest payoff. The buyer will rightly be concerned that while he may cooperate and make the purchase the seller could be rebelling, and ripping him off with prices far higher than what the product is worth. If this were true then instead of his cooperation giving him a payoff of 4 in place of 3, it would give a payoff of -1 instead of 0, which would involve a net loss.

To allay the buyer's fears the seller not only has to show that he gains from mutual cooperation in place of defecting and taking advantage of the buyer, as in the amended game matrix above, but he also needs to make a move in that direction to show that will be his choice. One possible method would be to offer store loyalty cards to build a lasting relationship with buyers, or to offer a generous returns policy in case the buyer was not happy with the product.

If the buyer does cooperate, based on the beliefs that the product is fantastic and the seller is cooperating too and not ripping him off, then there is still the issue of whether the seller actually cooperates too. The imagined game suggests he would but his real incentives are still to rebel as in the first game matrix in this section, as cooperation doesn't really give him a 4 payoff but a 2 payoff. He only pretended otherwise to encourage for a sale with the buyer. If he rips off the buyer then the interaction won't last long, but if the seller understands that buyers are essentially following a tit for tat strategy then a mutually cooperative outcome may be possible. That's as long as the buyer falls for the hype about the product and seller.

There is an alternative way for the mutually cooperative outcome to occur despite the incentives for players to rebel. That's with the seller instilling fear in the buyer. If done successfully that would see the payoffs a buyer gets from not buying the product falling, as he has been convinced that he can't do without it. The game matrix below shows the same payoffs as originally, except the two payoffs where a buyer rebels and doesn't buy the seller's product are now two less than before.

	Seller cooperates	Seller rebels
Buyer cooperates	2 2	3 -1
Buyer rebels	-1 1	0 -2

The seller's payoffs have not changed and there is no need to check those again, and his dominant strategy remains to rebel. But with the buyer's rebel payoffs considerably worse it's worth having another look at the choices from the buyer's point of view, assuming he falls for the lie that there's something missing in his life without the product.

Seller cooperates

I cooperate	2
I rebel	1

If the seller cooperates and sets a good price it's now in the best interests of the buyer to also cooperate. That gives him a payoff of 2 instead of the 1 from not buying the product from the seller.

Seller rebels

I cooperate	-1
I rebel	-2

If the seller rebels it's unbelievably still best for the buyer to cooperate and buy his product. That will give him a loss and a -1 payoff, but that's still better than the loss of -2 he'll get by not buying the product. The buyer has been convinced that he can't live without the product, and he'll tolerate any price set by the seller to get it.

The buyer now has a dominant strategy to cooperate, no matter what the buyer does. Combined with the seller's dominant strategy to rebel, a prediction can be made for the outcome of this game.

	Seller cooperates	Seller rebels
Buyer cooperates	2, 2	<u>3</u>, <u>-1</u>
Buyer rebels	-1, 1	0, -2

The Nash equilibrium of the buyer-seller interaction, assuming that the buyer falls for the seller's marketing campaign that he's currently in pain without the product, sees the buyer cooperate and buy the product for a -1 payoff, and the seller rebel and rip him off with an extortionate price for a 3 payoff. At this point neither player can (they believe) improve their payoff without the other changing their action. This is the seller's dream scenario, with him achieving the highest payoff possible. And as long as the buyer continues to believe the foolish notion that he must have the seller's product no matter what the terms, then the -1, 3 outcomes will remain the Nash equilibrium.

This game suggests that a seller looking for the highest payoffs shouldn't hype themselves and their products up directly, but focus a marketing campaign on the idea that buyers have something missing in their lives, and that the seller's product is the way to solve it. And as long as there is asymmetric information between the seller and the buyer there's no reason that this can't be achieved.

Fight or Flight

All of the games seen so far follow the form of the prisoners' dilemma, where players are expected to sabotage potential shared gains by following individual incentives to rebel, creating a Nash equilibrium result of mutual rebellion. The discussion has been over how players will try to change the game and dominant strategies to rebel, to move away from this outcome toward a better one. But there are many types of interaction that naturally follow a different pattern, and where the game is more adversarial in nature.

Evolutionary biology is one of the strongest supporters of game theory, and the rule of the animal kingdom is fight or flight. To adapt to this environment some creatures have evolved wings for fast movement that support flight to escape conflict, while others have grown strong teeth or claws to develop their ability to fight. But no animal will only ever be able to rely on just one strategy, and every animal must know when to fight and when to flight, with both options being possible outcomes in every game of interaction.

The game below is the well known hawk-dove game. It does not refer to the two different birds but instead to two different behaviours available to a player. There are two different players in this game and each has the choice to act as a hawk and fight, or to act as a dove and avoid fighting. The goal is to secure as much of a resource as possible, but to avoid a costly fight to get it. If both act hawk there is a fight before the two players share the resource, if both act dove they share the resource without a fight, and if one acts hawk and the other dove the former gets all of the resource and the latter nothing.

	Player 2 acts hawk	Player 2 acts dove
Player 1 acts hawk	$(V - C) / 2$ $(V - C) / 2$	0 V
Player 1 acts dove	V 0	$V / 2$ $V / 2$

In the game matrix above V = the value of a resource (money, food, a mate, etc.), and C = the cost of a fight. Unlike other games looked at this can't be solved right away by running through the best move for each player, as the values of V and C will depend on the specific circumstances. But different values for V and C can be tried out to assess the possible outcome options. The most important factor is whether V is higher or lower than C, and if V is higher then hawk behaviour and a fight may be worth it, but if V is lower than C then it may be best to avoid a fight and act as a dove. The cost of a fight will always be above 0, which ensures that the hawk-hawk and dove-dove outcomes will never be equal, and a choice will always have to be made.

First the game will be assessed in a world where a resource's value V is worth more than the cost of a fight C, with a value of V that is higher than C; $V = 4$, $C = 2$. That would see the payoff $(V-C)/2 = (4-2)/2 = 2/2 = 1$. And the payoff $V/2 = 4/2 = 2$.

	Player 2 acts hawk
Player 1 acts hawk	1
Player 1 acts dove	0

If player two acts as a hawk and fights then it's best for player one to do the same, for a payoff of 1 which is better than the 0 payoff from acting as a pacifist dove.

In the alternative scenario where player two acts as a pacifist dove and avoids a fight, the best response for player one is to act as a hawk. That earns a payoff of 4 in place of the 2 he'd get by acting as a dove, offering a return that's twice as high if payoff values were the same as they are here.

	Player 2 acts dove
Player 1 acts hawk	4
Player 1 acts dove	2

In a world where V exceeds C and the value of a resource is greater than the cost of a fight to get it, the dominant strategy for player 1 is to act as a hawk and fight, no matter what player two does. Because the payoffs are symmetrical in this game the same applies to player two, and both players will act as a hawk and fight to create a Nash equilibrium of hawk-hawk, and payoffs of 1, 1.

	Player 2 acts hawk	Player 2 acts dove
Player 1 acts hawk	<u>1</u> / 1	0 / 4
Player 1 acts dove	4 / 0	2 / 2

This is a prisoners' dilemma game, and when the value of a resource exceeds the cost to get it players will fight, even though both players could have achieved higher payoffs if they'd resisted the urge to do so.

Next the game will be assessed in a world where a resource's value V is worth less than the cost of a fight C, with a value of V that is lower than C; V = 2, C = 4. That would see the payoff (V-C)/2 = (2-4)/2 = -2/2 = -1. And the payoff V/2 = 2/2 = 1.

	Player 2 acts hawk
Player 1 acts hawk	-1
Player 1 acts dove	0

If player two acts as a hawk and fights then it's best for player one to act as a dove, for a payoff of 0 and not the loss of -1 that comes from a hawk-hawk interaction.

	Player 2 acts dove
Player 1 acts hawk	2
Player 1 acts dove	1

If player two acts as a dove to avoid a fight player one should act as a hawk, to take advantage of this. That would give a payoff of 2 and more than the 1 from a dove-dove interaction.

In a world where C, the cost of a fight for a resource, exceeds V, the value of that resource, there isn't a dominant strategy for player one but instead a mixed strategy. And as the payoffs remain symmetrical the same holds for player two here. Their highest payoff comes from doing the opposite of the other player, but they'd need to see what that is first before acting. Combining both players' payoffs gives the game matrix below.

	Player 2 acts hawk	Player 2 acts dove
Player 1 acts hawk	-1, -1	2, $\underline{0}$
Player 1 acts dove	$\underline{0}$, 2	1, 1

The long-run outcome here is for one player to act as a hawk and the other as a dove, and there are two underlined Nash equilibria. Once at that point neither player can improve their payoff without the actions of the other, and any attempt to improve their payoff will see a player worse off. This is the first game in this book where the Nash equilibrium has genuinely been for players to follow different roles instead of the same one, and without any lies or trickery this is the natural result. In every other game examined the long-run result has been for players to copy each other, whether to work together in cooperation or to do their own thing.

Looking at the issue of realistic values for V and C, the value of a resource and the cost of a fight to get it respectively, it seems likely that C would exceed V in most circumstances. The cost of a fight for all living things can be either death or serious injury, but there is always likely to be another chance to get the desired resource or a substitute to keep its value relatively low. This suggests that the second game matrix with players following different roles is the more realistic outcome in this hawk-dove game.

The result of the hawk-dove game supports evolutionary biology and a 'survival of the fittest' theory, where players find their own niche where they can fight best (hawk) and where others can't compete (dove), instead of all following the same path. It also supports the idea that the most important thing is not ability to fight, but the appearance of willingness to do so, which can put off the other player from acting hawk to avoid the cost of a fight. This might explain the popularity of bravado and posturing among both humans and animals in place of actual fighting. This strategy exploits asymmetric information among players about their willingness to fight, to get the gains offered by a fighting strategy without suffering the costs usually associated with it. The aim is to secure the hawk role and push the rival to select the dove role as a defensive measure.

The well known 'chicken' game looks specifically into the issue of bluffing rivals, and follows a similar format to hawk-dove game when the cost of a fight is greater than its prize. In this game two rivals face off, and their relationship could be defined by giving the name Enemy for player one and Opponent for player two. The two rivals sit in their cars, facing away from each other with an empty track between them. Then each driver starts their engine and forces a foot down on the thrusters, holding onto the steering wheel to keep the car straight on course on a path directed for the other player.

If both players in the chicken game were to stay on course as their cars accelerate forward then the cars will crash, for total carnage and negative payoffs of -10 each. If both drivers move and swerve then neither wins but neither loses anything either and the payoffs are 0 each. And if one flinches first and moves off course while the other stays, then the nervous mover suffers a -1 payoff by being shown as a coward who backed out of the duel, while the man who stayed the course gains a 1 payoff by virtue of successful one-upmanship over his rival. The following matrix puts this information into the game form.

	Opponent moves	Opponent stays
Enemy moves	0 0	<u>1</u> <u>-1</u>
Enemy stays	<u>-1</u> <u>1</u>	-10 -10

The likely outcome doesn't need to be calculated as the game is the same format as the hawk-dove example seen earlier. There are two Nash equilibria where each player follows a different strategy, underlined in the matrix above. At this point the mover wouldn't want to change his strategy, as then he would suffer a damaging crash and his negative payoff would become far worse. And the player who stays the course and doesn't swerve wouldn't want to change that, or he'd lose the payoff of 1 for the one-upmanship.

Like the hawk-dove game before it, the secret to getting the best payoff in this game is to show apparent willingness to stay the course no matter what the other player does (the rebel/fight/hawk option), and unwillingness to move the car and swerve (the cooperate/avoid fight/dove option), in order to get the other player to choose to move the car off the straight path. Stone cold nerves and lack of emotion would be a good signal that a player will do the first, while removing the steering wheel in front of the other player before the duel to make swerving impossible would serve the latter goal. But the actions must be believable to scare the other player, or the two rivals can find themselves facing a head-on crash that destroys them both.

Zero-Sum Games

The games above are all non-zero sum games, where it's possible for both parties to gain or lose together, even though their preferred individual outcome may be to see the other player lose as they gain. There has typically been a Nash equilibrium long-run outcome with rebellion by both players, although they'd be better off cooperating together, or alternatively two Nash equilibria where the long-run outcome may either be mutual cooperation or mutual rebellion. Even in the cases where the Nash equilibrium saw one player win and the other lose, there were still other options in the game that saw shared gains or losses, and it was a zero-sum outcome not a zero-sum game.

But there are games that are entirely zero-sum, where every possible outcome sees one player win and the other lose. The next game is such a situation, involving two players competing for sole ownership of a precious possession. In this game Fred starts with a ring on a chain around his neck, and both he and Callum want it all for themselves. If both players in this game follow the same strategy then Fred predicts what Callum will do and avoids his attempt to get the ring, keeping it safely around his neck. The payoffs would be 5 for Fred for holding onto the ring, and -5 for Callum for a failed attempt to claim it. But if the two follow different strategies then Callum catches him off guard, and he finally wins the ring for a precious payoff of 10, while Fred loses the ring that was his and that he's grown attached to for a terrible payoff of -10.

	Callum focuses on the ring	Callum ignores the ring
Fred focuses on the ring	-5 5	10 -10
Fred ignores the ring	10 -10	-5 5

More specifically, if both players focus on the ring then when Callum makes a grab for it Fred can keep it out of his grasp, and if both ignore the ring then if Callum attacks Fred directly he can defend himself. If Fred ignores the ring and Callum focuses on it, then the latter can pretend to attack and sneakily snatch the ring to get his prize. And if Callum ignores the ring and Fred focuses on it then the ring bearer is ripe to be ambushed, and as he obsesses over the ring Callum can approach from behind and knock him out to get his precious.

To predict what will happen we can look at things from each player's point of view, first from Fred's standpoint.

	Callum focuses on the ring
Fred focuses on the ring	5
Fred ignores the ring	-10

If Callum focuses on the ring Fred should copy him, as that offers a payoff of 5 instead of a -10 payoff loss.

	Callum ignores ring
Fred focuses on the ring	-10
Fred ignores the ring	5

If Callum instead ignores the ring for the time being then Fred should also ignore it, for a payoff of 5 instead of -10. Fred doesn't have a dominant strategy but a mixed one, and it involves doing the same as Callum does.

Next come Callum's own payoffs.

	Callum focuses on the ring	Callum ignores the ring
Fred focuses on the ring	-5	10

In the situation where Fred focuses on the ring Callum should ignore it, as that gives a payoff of 10 instead of the -5 payoff he'd get by also focusing on the ring.

	Callum focuses on the ring	Callum ignores the ring
Fred ignores the ring	10	-5

If Fred ignores the ring then Callum should do the opposite and focus on it, for a payoff of 10 in place of the -5 payoff he'd get by ignoring the ring like Fred. Callum also doesn't have a dominant strategy here but a mixed strategy; with his goal to do the opposite of what Fred does.

Putting all of these facts together shows that Fred gets his highest payoff when both players do the same thing, while Callum gets his highest payoff when the players do the opposite of each other. Or to put it more simply, ring bearer Fred wants to preserve the current dynamics between himself and Callum, while the creature without the ring wants to change the current dynamics of their interaction.

There isn't a long-run Nash equilibrium result here, where neither player can improve their own result without the other player, and in all of the four possible outcomes one player can change what they do for a better payoff. In zero-sum games like the one here there simply isn't a long-run outcome. Instead there are multiple short-run outcomes, which the two players may move between for short-run periods before the player on the losing end changes their strategy. The matrix below shows how the two players will move between different short-run outcomes, as they realize they can achieve a better payoff single-handedly by changing their individual strategy, without the other player having to change a thing.

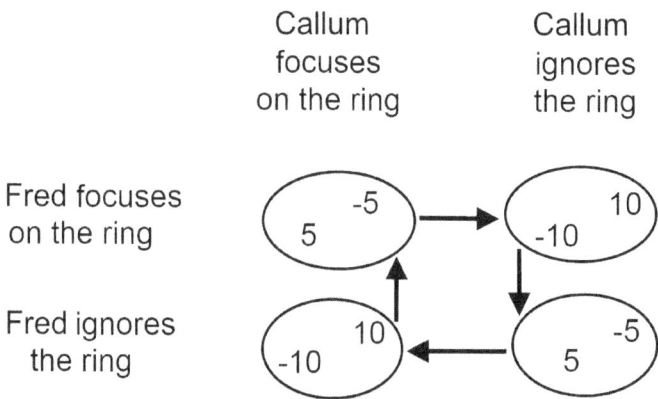

In the long-run in zero-sum interactions like this the game is likely to break down, as players grow tired of the lack of a stable outcome and push harder to achieve it, destroying either the other player or the prize they were chasing after.

When Game Theory Fails

Game theory has many critics who insist that the model is simply wrong, as its predictions do not come to fruition in the real world. One of the main complaints regards the fact that the model claims that people have individual incentives to betray others, and are likely to do so with regularity to prevent a superior collective outcome. But this isn't what we see in society as people cooperate every day at all levels.

Changing Games

In a prisoners' dilemma game individual incentives see prisoners betray each other every time as it's their dominant strategy, giving a Nash equilibrium long-run outcome of mutual rebellion where everyone does their own thing and care nothing for others. Going back to the first game in this book where guilty prisoners Bubba and Tyrone were being interviewed by police, game theory predicts each will incriminate the other in an attempt to either reduce the time they spend in jail, or eliminate it completely.

	Tyrone keeps quiet	Tyrone betrays
Bubba keeps quiet	-1 -1	0 -10
Bubba betrays	-10 0	<u>-5</u> <u>-5</u>

In this game the individual payoffs push both Bubba and Tyrone to betray the other, and give the outcome underlined where both men serve the five year sentence for their crimes. Yet there are countless examples where two guilty criminal suspects said nothing to police, and both were rewarded with a lower sentence as frustrated police could only charge them with a lesser crime. There are also many examples of one suspect taking the fall for the other, claiming that it was all down to them and that the other suspect is completely innocent, and willingly serving a longer sentence as a result. This seems to violate everything that game theory teaches.

But a game matrix only shows one type of interaction at a time, and the men involved may be thinking ahead to another completely different interaction away from police custody afterwards. Perhaps the suspects know they'll have to face the rest of the gang when they've left the police and served any prison sentence, and are acting in their best interests now with that long-run interaction in mind. The new game below shows suspect Bubba returning to face his gang, and having to face the consequences of his prior choice when questioned by police.

	Gang cooperates	Gang betrays
Bubba cooperates	10 10	-10 -100
Bubba betrays	-10 -50	-5 -100

If Bubba cooperated with the gang's wishes, keeping quiet about the crimes of the other gang member suspect Tyrone (whether Tyrone did or not), then when he left jail he would either receive cooperation from the grateful gang (payoff 10 for him), or the gang would betray and assassinate him anyway (payoff -100). But the gang would be more likely to keep on good terms with a reliable follower member by cooperating (payoff of 10 for them), than assassinate and lose one of their own who could be useful in the future (payoff -10).

In a situation where Bubba betrayed the gang and squealed on their illegal activities by implicating fellow member Tyrone, the gang would eventually track him down and make him pay. They may betray and kill off their useless gang member to see their numbers drop by one (payoff -5 for them), or instead cooperate with him to teach him the error of his ways and try to make him useful again, which requires effort and may be unlikely to work (payoff -10). Bubba would be assassinated if betrayed by the gang (payoff -100 for him), and even if they cooperated and kept him alive he would still be tormented as they try to teach him a lesson (payoff -50).

The dominant strategy for the gang is to do whatever Bubba did, and they can be expected to match his cooperation or betrayal. For Bubba, the strategy to cooperate weakly dominates the strategy to betray, and if the gang was to betray him the strategy offers no worse a payoff, but if the gang were to cooperate with him this choice would offer a far better outcome. His only option is therefore to cooperate and hope the gang doesn't betray him, and he can do no more with the payoffs on offer. But the incentives the gang face should see this strategy save Bubba, unless they too face another game of a different nature after this one, and then anything is possible.

With this additional game taking place after the prisoners' dilemma game earlier, it can shed light on an unexpected decision by Bubba to keep quiet and cooperate when the two suspects were being questioned by police. It isn't that game theory is necessarily wrong, but that the particular game hasn't accounted for additional games of a different kind to come in the future.

Altruism

Another argument used to discredit game theory is the existence of altruism, and people putting the interests of others before their own. This goes against everything the model teaches about players following their individual incentives and payoffs. The game matrix below is a replication of the prisoners' dilemma game seen earlier between Man and Woman, and it will be used to test the effects of altruism.

	Woman cooperates	Woman rebels
Man cooperates	2 2	3 -1
Man rebels	-1 3	<u>1</u> <u>1</u>

The model predicts that both players will rebel against the other and the underlined outcome is what we would expect to occur, with payoffs of 1 each. If both players adopted a tit for tat strategy then the cooperative outcome in the top left may also be possible in the long-run, with payoffs of 2 each. And if this game was only a short-run interaction to be followed by a game of a different kind, as just seen when Bubba returned to his gang, an outcome of one player cooperating as the other rebels is also possible.

But as far as game theory is concerned the two outcomes where one player cooperates and one rebels simply isn't possible in the long-run interaction of an unchanging game, as neither is a stable Nash equilibrium where neither player can improve their payoff single-handedly. In a situation where Woman cooperates and Man rebels, Woman will get a negative payoff of -1, but she could change it at any time by rebelling for a better payoff of 1, and the model assumes that she always will.

But there are many long-term relationships where a man or woman rebels relentlessly, taking more and more without ever giving anything in return, yet the other partner may continue to cooperate and altruistically tolerate their flaws. The game theory model doesn't appear to model this scenario, and one payoff needs to be changed in the game above to make it represent altruism.

	Woman cooperates	Woman rebels
Man cooperates	2 2	3 -1
Man rebels	$\underline{3}$ $\underline{3}$	1 1

In this amended game matrix all payoffs are the same as before, except now the scenario where Man rebels and Woman cooperates has a different payoff for Woman. It used to be a negative payoff of -1, and is now a positive payoff of 3. If this payoff was accurate then it could explain why a rebellion/cooperation relationship can be maintained for the long-term, and why the cooperating player doesn't simply rebel like the other. In this game the dominant strategy of Man is still to rebel, but Woman's strategy is now no longer to rebel but to do the exact opposite of what Man does. Combining the two strategies would give a stable long-run Nash equilibrium outcome of Man rebels and Woman cooperates, for payoffs of 3 each.

The only question is whether this new payoff of 3 for Woman in the bottom left pairing of the grid is more accurate than the -1 payoff she had before, when Man still gets a -1 payoff for cooperating with a rebelling player in the top right pairing of the grid. The difference could possibly be explained as follows: when Man attempts to cooperate with a rebelling player he struggles and suffers a loss, but when Woman in this example attempts to cooperate with a rebelling player she simply altruistically accepts Man's payoffs as her own. Man achieves a payoff of 3 so as far as Woman sees it she also achieves a payoff of 3, as she has forgotten about her own wants and evaluates her payoffs solely upon the effect they have on the other person.

This can explain unexpected long-run outcomes where one player appears to accept suffering a poor payoff over an extended period, despite having the power to individually change it. It isn't game theory failing to represent the situation but incorrect payoffs being entered into the game matrix, which don't represent the altruistic situation taking place.

Hatred

One final issue accusation levelled at game theory is that outcomes that should be impossible do come to pass. The game below is the chicken game seen earlier, where two cars square off in a face to face duel to see who twitches first. Player one is referred to as Enemy and player two is known as Opponent.

	Opponent moves	Opponent stays
Enemy moves	0 / 0	<u>1</u> / <u>-1</u>
Enemy stays	<u>-1</u> / <u>1</u>	-10 / -10

In this chicken game the two Nash equilibria were where one player stays the course and the other moves, with payoffs of 1 and -1 respectively. At this point neither player can singlehandedly achieve a better payoff. If players were nervous about the threat of a collision and the -10 payoff then it's feasible that both could move at the last second, but according to game theory's assumptions the outcome where both players stay the course and crash shouldn't occur. Yet there are many examples of two rivals destroying each other, whether in a duel by car, gunfight, or various other means.

The issue of hatred may be able to explain the failure of game theory's predictions in a confrontational game such as chicken. In every game seen so far, from the prisoners' dilemma, to repeated, mixed strategy, or zero-sum games, the implicit assumption has been that players want to maximize their own payoffs. But this may not always be true. At times players may be more concerned that others don't gain, and in an interaction with a mortal enemy the most important goal for a player may not to secure their own high payoff, but to force a low payoff on their rival.

In the game earlier players tried to act like they would never move but were determined to stay the course, with the intention to make the other player believe that they had to be the one to move, or suffer the disaster and injury of a head-on collision. One player moving as the other stays the course is the Nash equilibrium and long-run outcome, and what both players would expect to occur. But there's no reason why a hateful player can't take advantage of all this and secretly work toward their own agenda, as the other player thinks that it's their agenda being followed and naturally assumes their rival will be chasing his own higher payoff.

If the player 'Enemy' was so filled with hate that his only goal was the destruction of the other player 'Opponent', then he would surely be willing to sacrifice his own goals to ensure this outcome was achieved. Enemy's goal would be to see Opponent suffer the -10 payoff from a head-on collision, and he wouldn't care if he suffered the same fate himself. It should actually be quite easy to achieve this goal, as all the hateful player needs to do is stay the course and get his rival to stay put too, which he would naturally want to do. If the hateful player Enemy showed some signs of weakness before the race, or simply didn't show signs of strength, then that could convince the overconfident player Opponent that his rival would move the car and swerve, even if it was at the last moment. But that time would never come and instead both players would suffer a crash, just as the bitter player Enemy had intended and planned.

The phenomenon of hatred causing unpredictable and suboptimal results can be seen far beyond the chicken game. Those sitting in jail for murder are not usually there due to a rational decision to follow their own high payoffs, but an emotional decision to force a low payoff on another person. This is perhaps the one area that game theory fails to represent, and the model always assumes rational decision making and not acting on emotion. And when players don't act rationally and in their own best interest then anything is possible, and game theory's ability to predict future outcomes may be limited.

www.ingramcontent.com/pod-product-compliance
Lightning Source LLC
Chambersburg PA
CBHW051732170526
45167CB00002B/908